Water
74

矿井水
Mine Water

Gunter Pauli

[比] 冈特·鲍利　著

[哥伦] 凯瑟琳娜·巴赫　绘

郭光普　译

上海远东出版社

丛书编委会

主　任：田成川

副主任：何家振　闫世东　林　玉

委　员：李原原　翟致信　靳增江　史国鹏　梁雅丽

　　　　任泽林　陈　卫　薛　梅　王　岢　郑循如

　　　　彭　勇　王梦雨

特别感谢以下热心人士对童书工作的支持：

匡志强　宋小华　解　东　厉　云　李　婧　庞英元

李　阳　刘　丹　冯家宝　熊彩虹　罗淑怡　旷　婉

杨　荣　刘学振　何圣霖　廖清州　谭燕宁　王　征

李　杰　韦小宏　欧　亮　陈强林　陈　果　寿颖慧

罗　佳　傅　俊　白永喆　戴　虹

目录

Contents

一个凉爽的下午，一群鹌鹑正在一片金黄的油菜花田里享用晚餐。

"我们在你这里收获了很多，为什么你开的花比以往更亮丽呢？"一只鹌鹑问道。

"这要感谢矿井水。"油菜花答道。

It is a crisp spring afternoon. Some quail are enjoying their dinner in a field of beautiful yellow rapeseed flowers.

"We are getting a great harvest from you. Why are you blossoming brighter than ever before?" asks one of the quail.

"It's thanks to the mine water," responds the rapeseed flower.

比以往更亮丽

Blossoming brighter than ever before

As deep as four thousand metres

"矿井水？我知
道雨水与河水，还有从大坝
和地下来的水，但矿井水是什么？
对我来说可真新鲜！"
"你知道这里的矿井有多深吗？"
"当然，钻机已经往地下打了几
十年了，有些都打到4 000
米深了！"

"Mine water?
I know about rainwater
and river water. And there's
also water from dams and
aquifers, but mine water? That's
new to me."
"Do you know how deep the mines
are around here?"
"Yes, the shafts were dug over
decades and some may be as
deep as four thousand
metres."

"什么？都快打到地狱了！"油菜花惊讶地说。

"噢！即使温度可能上升到50℃，那里也没有地狱。但是你能想象要乘多久电梯才能到那里吗？"鹌鹑问道。

"What? That's getting close to hell!" exclaims the flower.

"Well, even though the temperature could rise to fifty degrees Celsius, of course there is no real hell there. But can you imagine how long it must take to get down there in a mine elevator?" asks the quail.

想象要多久才能到那里

Imagine how long it must take to get there

卡车在地球的肚子里开来开去?

Trucks driving around in the belly of the Earth?

"我想肯定很长！

而且在这么深的地下工作肯定

每天都像在洗桑拿！"

"我同意，那真是一种挑战！人们采矿时，

机器负责降温。在地下工作也不再像以前那样

艰难了。还有卡车和火车在地球的肚子里运

输物品。"

"卡车在地球的肚子里开来开去？那

排出的废气怎么办呢？"油

菜花问道。

"Quite
long, I suppose. But
working so deep underground
must be like being in a sauna all
day."

"It's quite a challenge, I agree. While
people are taking out the ore, machines
cool the air though. Working underground is
not as hard as it used to be either. There are
even trucks and a train for transport inside
the belly of the Earth."

"Trucks driving around in the belly
of the Earth? What happens to all
the exhaust fumes?" asks
the flower.

"为了换气，矿井里有世界上最精密的空调系统，就安装在我们脚下几千米的地方。"

"唯一的问题是当矿井关闭的时候，水会流进去。时间长了，所有东西都会坍塌。"

"不过，矿井里的水肯定会被泵出去吧？"鹌鹑说，"你想想，这可是地球内部的纯净水呀！"

"To freshen the air, mines have some of the most elaborate air conditioning systems in the world, located only a few kilometres under our feet."

"The only problem is that when a mine is closed down, water flows into it. Over time everything could collapse."

"But surely mines have to pump the water out?" remarks the quail. "Imagine, pure water from the inner sanctum of the earth."

世界上最精密的空调系统

Most elaborate air conditioning systems

How to properly manage the waste

"噢，是的，从地球深处来的水不仅纯净，还很温暖。它能让我的根保持温暖，所以冬天我也能很容易生长在矿井附近的田地里，而这些地没人要。"

"人们为什么不愿意生活在这里呢？"

"因为以前矿井周围都是人们扔掉的垃圾。那时，人们还不知道如何正确处理这些从地下挖出的废物。"油菜花解释道。

"Oh yes, the water that comes from deep, deep down is not only pure, but also warm. It helps to keep my roots warm so I can easily grow throughout the winter in the fields around the mines – on land that no one wants."

"Why don't people want to live on that land?"

"Because before, a lot of rubbish was just left lying around the mines. Back then, people didn't know how to properly manage the waste that comes out of the earth," explains the flower.

"现在，我们可以用废弃的岩石和沙子造纸，还可以用抽上来的水灌溉农田。但是，我不知道这些水是否安全。水里是不是含了矿井里的一些化学物质？"

"这些水除了岩石和金子，从没和其他东西接触过。水里甚至可能会有几片小金片呢！"

"难怪你们有这么金灿灿的颜色！"鹌鹑大喊道。

"These days, we make paper from the rocks and sand that's left behind and we know that we could use the water for irrigation. But, I wonder if this water is safe for us to drink. Isn't it contaminated with chemicals from the mine?"

"This water has never been in contact with anything except rocks and gold. It could even have a few tiny flakes of gold in it!"

"No wonder you rapeseeds have such a bright golden colour!" exclaims the quail.

这么金灿灿的颜色!

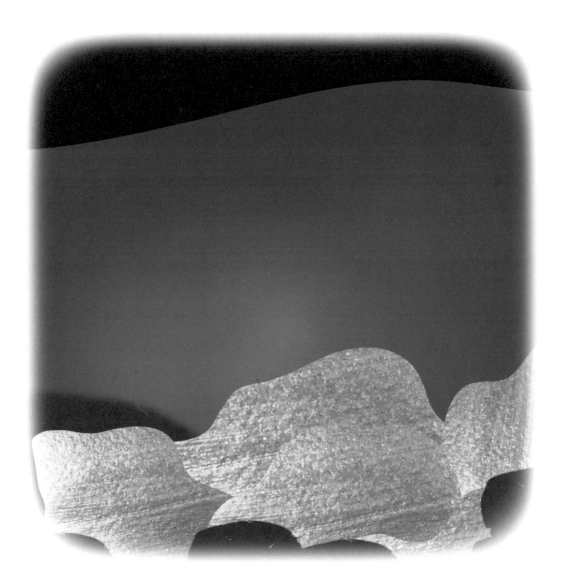

Such a bright golden colour!

用来做生物燃料或蜡烛

生物燃料

Used as biofuel and to make candles

"我知道我们鹌鹑是为数不多的可以安全地吃油菜籽的动物。我还知道人类和其他动物不会吃长在这里的庄稼。所以，有时候我在想我们是否应该冒这个险……"

"不用担心，对你来说这是非常安全的！"

"你知道吗？从生长在矿井附近的农作物中提取的油可以用来做生物燃料和蜡烛。"油菜花问道。

"I know that us quail are some of the only animals that can safely eat rapeseed. And I know that people and animals should not eat crops that are grown here. So, I sometimes wonder if we should take the risk …"

"Don't worry, it's perfectly safe for you!"

"And did you know that oil from the crops grown near the mines can be used as biofuel and to make candles?" asks the flower.

"这太棒了！"小鹌鹑喊道。

"我在想你的油是不是可以用来做唇膏，让女人们把自己打扮得更漂亮。"

……这仅仅是开始！……

"That's fantastic!" exclaims the quail.
"Now I wonder if your oil can be used in the lipstick that women could use to make themselves even more beautiful."

... AND IT HAS ONLY JUST BEGUN!...

AND IT HAS ONLY JUST BEGUN!

Did You Know ?

你知道吗？

The deepest mine in the world is 3,9 km deep and is located to the southwest of Johannesburg, South Africa. Of the 10 deepest mines in the world, 8 are found in the same region. The other two are in Ontario, Canada.

世界上最深的矿井在南非约翰内斯堡的西南部，有 3 900 米深。世界上最深的 10 个矿井中的 8 个都在这个地区，另外两个在加拿大的安大略省。

The deeper the mine, the hotter it gets down below and also the more costly it becomes to pump fresh air into the shafts. This fact drives the mining industry towards automation.

越往矿井深处就越热，把新鲜空气打到矿井里的费用也越高。这一事实推动了采矿工业的自动化。

The first steam-powered engine to perform mechanical work was the Newcomen engine, designed by Thomas Newcomen in 1712. Hundreds of these engines were built in the 18th century and were used to pump water out of mines.

第一个进行机械工作的蒸汽动力发动机是 1712 年由托马斯·纽卡门设计的纽卡门发动机。18 世纪人们制造了数百台这样的发动机把水从矿井里抽出来。

Rapeseed, also called canola (after one of the varieties), is the third largest source of vegetable oil. In China and Southern Africa, the rapeseed plant is eaten as a vegetable. The oil used to be used as a bio-lubricant for steam engines.

油菜籽也被叫作芥菜籽，是植物油的第三大来源。在中国和南非，油菜还可以作为蔬菜食用。菜籽油曾经被用作蒸汽发动机的润滑油。

The rapeseed plant gets its name from the Latin name for 'turnip', i.e. rapa. For marketing purposes a new name was created: canola, which is a contraction of Canada and oil low acid.

油菜的英文名 rapeseed 来源于芜菁的拉丁名 rapa。出于商业目的，人们发明了一个新名字：canola。这个词是"加拿大"和"低酸油"的缩略。

One tonne of rapeseed yields 400 litres of oil. It takes the seeds 150 days to mature.

1 吨油菜籽可以榨出 400 升油。油菜籽需要生长 150 天才能成熟。

$\mathcal{N}$early 70% of all accessible fresh water is used in agriculture – that is more than triple that used for industrial purposes (23%). Only 8% is used for municipal use.

人们把近 70% 的可用淡水都用在了农业上——相当于工业用水（23%）的 3 倍多，只有 8% 是市政用水。

$\mathcal{Q}$uail are hardy, disease-resistant birds. Hens lay 270-300 eggs per year, for three to four years. The quail egg has been described as a mineral cocktail, containing many of the essential minerals people require.

鹌鹑是适应力强又抗病的鸟类。母鹌鹑每年可以产 270—300 个蛋，而且可以产三四年。鹌鹑蛋被誉为"矿物质鸡尾酒"，含有许多人体必需的矿物质。

Think about It

想一想

Would you like taking an elevator that drops you four kilometres down a mineshaft?

你愿意乘坐电梯到达 4 千米深的矿井吗?

How would it be not to know if it is morning, noon or night, or whether it is raining or hot outside?

如果分不清现在是早晨、中午还是晚上，也不知道外面是否下雨或炎热，你会有什么感觉?

你会吃种在原来是矿场的土地上，现在又用矿井里的水灌溉的食物吗?

Would you be prepared to eat food grown on land that was previously mined and is now irrigated with water from a mine?

When you irrigate plants on contaminated land with pure water from a mine, could it be possible to purify the soil over time, distributing the pollutants in very small amounts?

如果长时间地用矿井里的纯净水浇灌种在污染的土地上的植物，是否可能把污染物含量降低，进而净化土壤?

Do It Yourself!

自己动手！

The colour yellow symbolises energy, positivity and hope. Do a test to see if yellow does indeed have a positive, energising effect on you, your friends and your family. Collect pictures of yellow flowers such as sunflowers or rapeseed flowers as well as pictures of blue and purple flowers, like the gentian. Show them to others and ask them how looking at the different coloured flowers makes them feel. Compare your feedback with what your friends got when they did the same.

黄色代表了能量、积极和希望。做一个测试，看看黄色是否真的对你、你的朋友和你的家人有积极的、激励的效果。搜集黄色花朵的照片，比如向日葵或油菜花，同时也搜集蓝色和紫色花朵的照片，比如龙胆。让其他人看看这些照片，询问他们看到不同照片时的感觉。比较一下你的感受和朋友们的反馈。

学科知识
Academic Knowledge

生物学	油菜籽可以产出大量花蜜；鹌鹑具有强大的免疫系统，不易患病，所以不需要抗生素来治疗；小鹌鹑刚孵出后吃虫子，5—6周以后就只吃植物了；人工修建的湿地在一定程度上可以修复酸性废水的污染。
化 学	油菜籽含有50%的芥酸；种植油菜需要高浓度的氮肥；矿井里的酸废水是由大量的天然硫矿，尤其是黄铁矿的氧化造成的；当氧化反应发生时，氢离子被释放出来，pH值就降低了；用石灰、硅酸钙或者碳酸盐中和的方式恢复pH值。
物 理	托马斯·纽卡门设计的空气发动机。
工程学	采矿工程是一门综合学科，包括地质学、测绘学、选矿学、冶金学以及岩土工程学；芬兰桑拿浴和土耳其浴在结构上的区别。
经济学	利用现有的设备在现有的矿中获得更多的资源可以减少资本投入；投资采矿业最重要的不可知因素是停止开采后的花费，即使该地区恢复到原来的生态状态的费用；碎石头（废弃物）可以用来制造纸张；金属元素是可以100%循环使用的。
伦理学	采矿是危险的职业，所以在古埃及和古罗马时期都是由罪犯和战俘来做的；国际公约中已经禁止强迫劳动。
历 史	詹姆斯·瓦特设计的（蒸汽）发动机是托马斯·纽卡门设计的（空气）发动机的改进版，只要用一半的燃料就能产生相同的动力；大约4 000年前埃及的沉井就达到了地下90米深，罗马的矿井则达到了200米深；在炸药发明之前，矿工用火力碎石；桑拿产生于拜占庭帝国，并被斯拉夫商人带到了芬兰。
地 理	格陵兰和纽芬兰有最古老的桑拿；南非的约翰内斯堡是围绕金矿而建的，也是世界上少数几个不是建立在森林或大河附近的主要工业城市之一。
数 学	纽卡门发动机的工作频率为每分钟12次，每次能够泵出数十升水；鹌鹑每天需要喂食20—25克，而鸡需要120—130克，而且鹌鹑在5—8周后就能达到上市体重（140—180克），之后将生产200—300只鹌鹑蛋。
生活方式	健康的生活方式包括卫生和保健：从外部和内部保持身体清洁。
社会学	意大利诗人但丁写了一部《神曲》，他对地狱毛骨悚然的描写已经成为西方思想中根深蒂固的地狱形象，并且赋予了米开朗基罗、弥尔顿和艾略特等人创作的灵感；"桑拿"在芬兰语中是指浴室。
心理学	黄色是使人振奋的颜色，可以使人们更乐观、自信、自尊和有创造力，尤其是绽放着油菜花和向日葵的黄色田野。
系统论	现代文明需要矿山、矿石和金属来驱动，然而这经常会对环境造成过度危害；由于采矿的社会成本也很高，所以现存的经营模式需要重新设计。

情感智慧
Emotional Intelligence

小鹌鹑

小鹌鹑知道油菜花是他们的食物来源之一，他赞美油菜花，并且还问她们为什么比以前长得更好看、更鲜艳。小鹌鹑对水很好奇，他问了很多问题以期能发现更多。小鹌鹑和油菜花的对话就是信息的交换。小鹌鹑愿意回答所有问题并无偿地分享他的知识，包括一些实用的方法，比如用石头造纸。然而，小鹌鹑担心矿井水的质量，并勇担责任，建议矿井水不应该用于生产供人类食用的粮食。小鹌鹑了解到用菜籽油做蜡烛和生物燃料是安全的，但他想知道菜籽油是否也可以用来做唇膏。

油菜花

油菜花很谦虚。当她收到赞美时马上把功劳归于纯净的矿井水。油菜花是无知的，但她很爱学习，所以她用自己的问题来回应小鹌鹑的问题。油菜花喜欢利用矿井水，但她很惊讶矿井竟然这么深，人们不得不在那么闷热的条件下工作。小鹌鹑解释了空气是如何保持新鲜的。油菜花立刻形成了自己的观点，并认识到关闭矿井是个问题，因为里面的水还得抽出来。油菜重新整合了已知的所有信息，并思考如何利用矿井水度过寒冷的冬天。油菜花的金黄色代表了乐观和自尊。即使鹌鹑提出了警告，油菜花依然保持乐观，并注意到了使消极变成积极的机会。

艺术
The Arts

你能想象黑暗中的生活是什么样子吗？想像你在地下4 000米的地方，很闷热，水的温度已经上升到了威胁你的生存的程度。听起来像是恐怖小说的开头！你能不能以你深入到地下数千米开头，写一篇惊悚故事？

思维拓展
Systems: Making the Connections

开矿和做外科手术的逻辑是一样的，要多为病人的健康考虑，尽量不留下不良后果。地下采矿比露天开采导致的危害要少得多，但产生的酸性矿井水却是严重的环境威胁。采矿工程能够按照最严格的健康、安全和环境标准进行。然而，黄铁矿的出现推动了化学采矿的应用，不断产生的污染会毁坏水路、土地，对地球和人类的健康造成了很大的风险。同时，裂隙水的出现威胁到了矿井和地面的安全（地下水会淹没矿井）。这种水不一定受到了污染，可能很纯净，也可以饮用，但从矿井安全的角度看，必须把水抽出来。这不仅是一种很大的能源耗费（尤其是从几千米的矿井下抽到地面），也使矿主承担了很大的责任，要为矿井关闭后的水处理负责。在干旱的地区突然降下了大量的水，既带来了机会，也带来了挑战。矿井周围的地区遭受着污染物导致的严重污染，如放置碎石的尾矿坝形成的粉尘。另一个主要污染源是依赖有毒化学反应的处理技术。结果，本来可以用来灌溉土地的营养水现在却被污染了，只能用来灌溉生产蜡烛用油和生物燃料的农作物，而不适合被人类食用。把水和土地结合起来生产能作为燃料的农作物，有很多好处。第一，燃料可以立刻使用；第二，油可以从种子中提取（每吨可以生产400升），而废弃的、不能喂动物的油渣饼现在经处理后可以生产沼气。如此将进一步增加能源产出。灌溉和燃料生产的连续过程可以慢慢减少土地的污染并使土地恢复到原本的状态，化负面效果为正面效果：生产燃料、清洁土地、增加就业，这将弥补采矿自动化造成的失业，而自动化对在困难条件下开采越来越深的岩矿是很有必要的。

动手能力
Capacity to Implement

你见过鹌鹑蛋吗？去食品店或商场买一些这种小小的蛋，给人们看看并请他们告诉你鹌鹑蛋和鸡蛋的区别。除了大小不一样，你的朋友和家人还能告诉你这两种蛋和这两种动物有什么不同吗？记下你学到的内容，然后作出明智的决定：你要吃哪种蛋？为什么？

故事灵感来自
This Fable Is Inspired by

吉尔·马库斯
Gill Marcus

吉尔·马库斯的祖辈是移民到南非的立陶宛人。由于她的父母是反种族隔离的积极分子，她在 1969 年便和兄弟姐妹跟着父母流亡海外。1976 年她获得了南非大学工业心理学的商学士学位。加入非洲国民大会以后，她开始在它设在伦敦的办公室工作。她于 1994 年当选为议员，并在 1996—1999 年任财政部副部长，1996—2009 年任储备银行副行长。她还担任了西部矿业公司的总裁，后来成为金田有限公司的董事会成员，并敦促管理层考虑在实践中进行改革。

图书在版编目（CIP）数据

冈特生态童书.第三辑修订版：全36册：汉英对照 /
(比)冈特·鲍利著；(哥伦)凯瑟琳娜·巴赫绘；
何家振等译.—上海：上海远东出版社,2022
书名原文：Gunter's Fables
ISBN 978-7-5476-1850-9

Ⅰ.①冈… Ⅱ.①冈… ②凯… ③何… Ⅲ.①生态环
境-环境保护-儿童读物—汉、英 Ⅳ.①X171.1-49

中国版本图书馆CIP数据核字(2022)第163904号
著作权合同登记号图字09-2022-0637号

策 划 张 蓉
责任编辑 祁东城
封面设计 魏 来 李 廉

冈特生态童书
矿井水
[比]冈特·鲍利 著
[哥伦]凯瑟琳娜·巴赫 绘
郭光普 译

记得要和身边的小朋友分享环保知识哦！
八喜冰淇淋祝你成为环保小使者！